It’s a piece of code.
Anyone can do it.

Published by The Honey Bee Company LLC
2906 Central Street • Suite 258
Evanston, IL 60201
847.864.8032

ISBN 978-1-947933-06-4

My name is

____

I’m ready to get to work!

It seems that grownups are always telling
Paige and Paxton what to do.

What to wear.

When to go to bed.

What to eat.

When to do chores.

Match the phrases to the correct pictures.

Grandmama Puzzle says today will be different.

Count the yellow flowers.

There are ________ yellow flowers.

Grandmama Puzzle says when they get to the hackathon, they'll get to tell computers what to do.

Color the picture.

If anyone knows about computers
it's Paige and Paxton's grandmother.

Circle the picture that's different.

She started working on computers right after college.

Circle the pictures that are the same.

Now she teaches others about computers.

Count the computers on this page.

Now count the computers on this page.

There are ________ computers all together.

That's why the Puzzles are going to a hackathon.

[redacted]

[redacted]

At the hackathon, Paige and Paxton will see computer programmers like their grandmother working on computers.

____

____

Look at the pictures on both pages.
What happened first, second, third and last?
Number the order of events.

Computer programmers are at the forefront of technology.

Circle the one thing that doesn’t belong.

Technology is creating new things and new ways of doing things to make our lives easier.

Draw lines to the ones that are the same.

Computer programmers love to come up with all kinds of new ways for computers to make our lives easier.

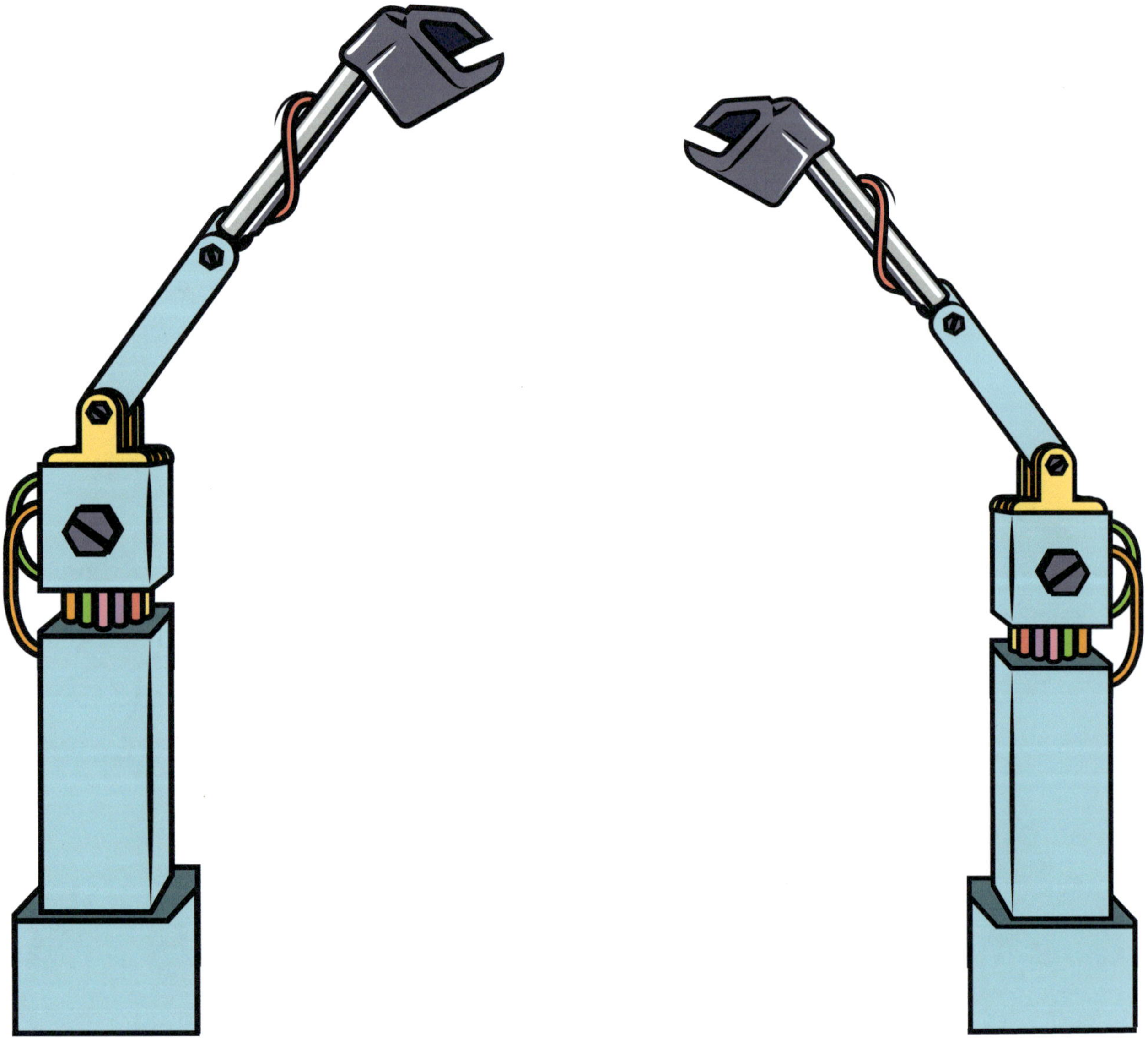

Circle the shorter robotic arm.

Measure the taller glass.

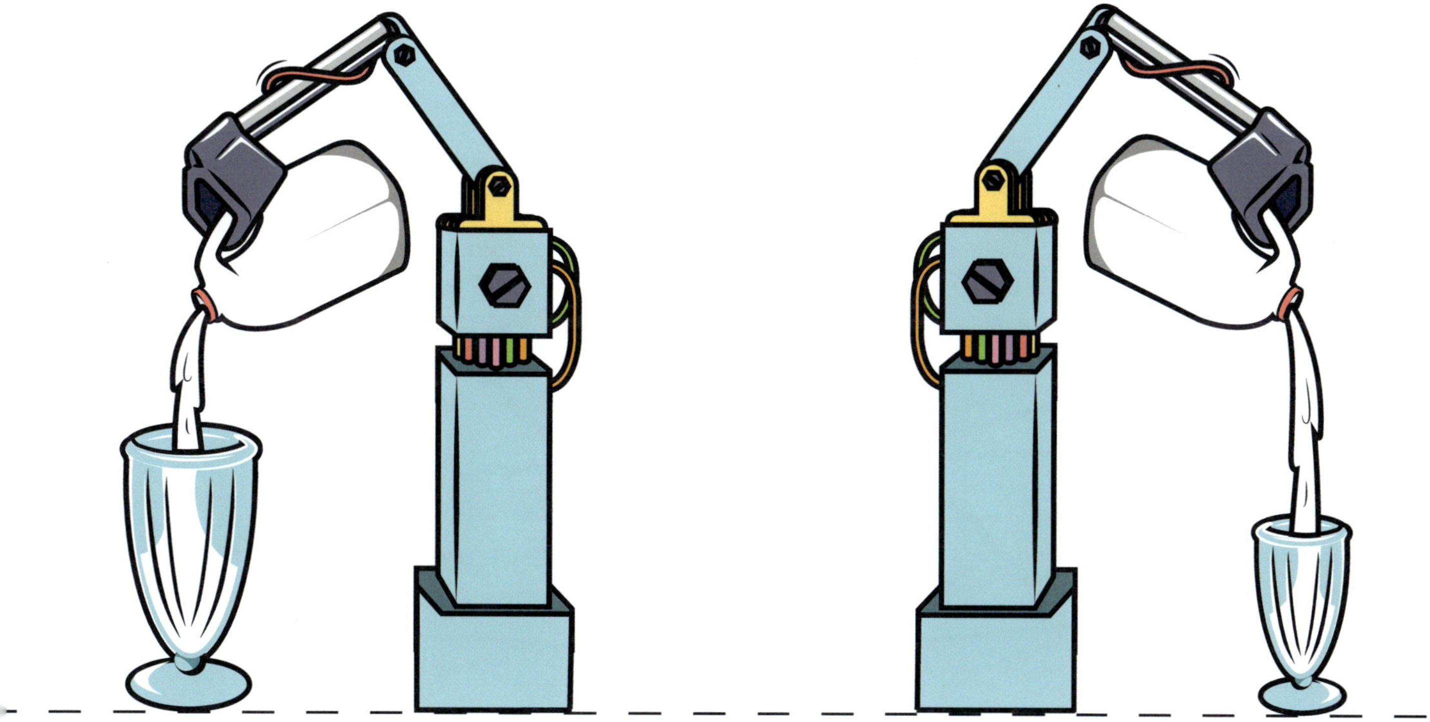

The taller glass is ________ inches high.

Computer programmers make it all possible by writing code, the instructions that tell computers what to do.

Color the picture.

At first Paige and Paxton are horrified that computers are like kids: always being told what to do.

Draw a picture of Paige and Paxton horrified.

But being told what to do has some benefits.

If no one told computers what to do,
they wouldn't know what to do.

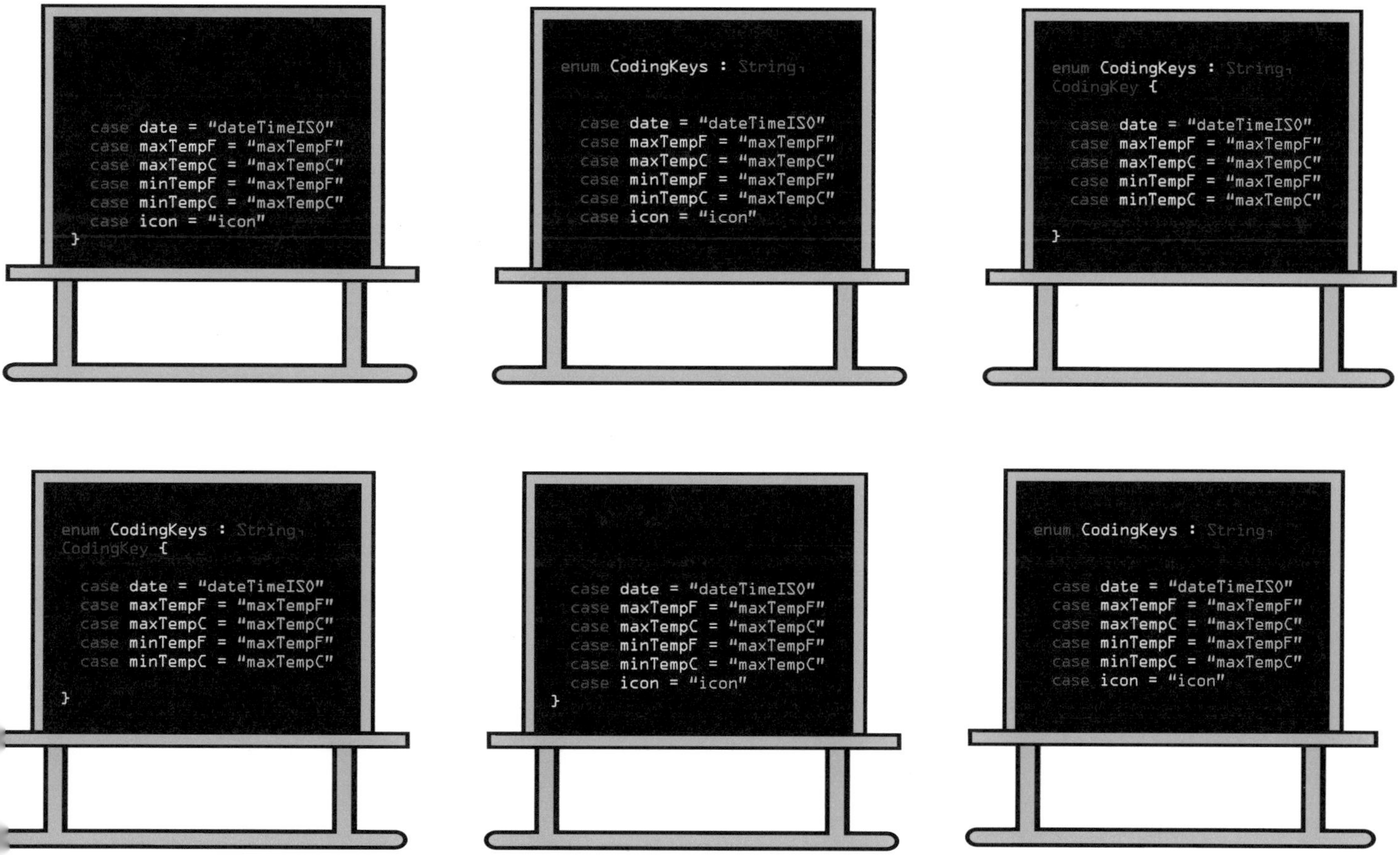

Draw lines to the ones that are the same.

Computers don’t know we want money from an ATM.

Look at the pictures. What happened first, second and last? Number the order of events.

Computers don’t know we want to pay for a cup of coffee.

Number the order of events on this page too.

That’s why we need computer programmers.

End

15	___	9	___	3
___	11	___	5	___
13	___	___	___	1

Start ↑

Fill in the missing numbers to help
the computer programmer get to her computer.

It’s their job to write the code that tells computers what to do and when to do it.

Color the picture.

Now Paige and Paxton know that being told what to do is a good thing for computers and for them.

Telling computers what to do is called

Programming

Telling computers what to do is also called

_____ _____ _____ _____ _____ _____

Telling kids what to do is called

Guidance

Grandmama Puzzle says guidance makes sure we do the right things and move in the right direction.

Count all the napkins.

There are ________ napkins.

With a little guidance, Paige and Paxton can learn to tell computers what to do.

You can too.
What would you like to tell computers to do?

Draw a picture.

Our Mission

Change the course of a nation by making STEM (Science, Technology, Engineering and Math) easy to teach and easy to learn at the very beginning of a child's educational experience

Our Vision

World Domination through STEM

Why We Do What We Do

For Kids: 80% of adults in STEM fields had at least one parent in the field. This means a huge number of our kids are missing out on opportunities that can be life changing.

For Teachers: Classroom teachers face unprecedented challenges to meet expectations and standards. Teachers need time-saving resources that enable them to make small shifts in their classroom lessons to drive STEM interest and create a STEM mindset.

For Our Nation: We're losing the battle. Thirty years ago, 40% of the world's scientists and engineers came from the U.S. Today, it's 15%. STEM fields are where the well paying jobs are—and will be—in the global economy.

Like Us On Facebook

Pin Us On Pinterest

Follow Us On Twitter

Made in the USA
Monee, IL
05 November 2026

67380374R00021